Dilermando Dourado Pacheco
Danley Freitas Gonçalves
Rafael Tálison Magalhães Campos

Acidification of soils with the use of S and cattle manure in northern Minas Gerais

Dilermando Dourado Pacheco
Danley Freitas Gonçalves
Rafael Tálison Magalhães Campos

Acidification of soils with the use of S and cattle manure in northern Minas Gerais

Analysis of the responses of pH in water, SMP pH, H+Al, EC, Ca and Mg to the treatments

Imprint

Any brand names and product names mentioned in this book are subject to trademark, brand or patent protection and are trademarks or registered trademarks of their respective holders. The use of brand names, product names, common names, trade names, product descriptions etc. even without a particular marking in this work is in no way to be construed to mean that such names may be regarded as unrestricted in respect of trademark and brand protection legislation and could thus be used by anyone.

Cover image: www.ingimage.com

This book is a translation from the original published under ISBN 978-613-9-70281-7.

Publisher:
Sciencia Scripts
is a trademark of
Dodo Books Indian Ocean Ltd. and OmniScriptum S.R.L publishing group

120 High Road, East Finchley, London, N2 9ED, United Kingdom
Str. Armeneasca 28/1, office 1, Chisinau MD-2012, Republic of Moldova, Europe
Printed at: see last page
ISBN: 978-620-8-18195-6

Authors:

Dilermando Dourado Pacheco

PhD in Plant Science from the Federal University of Viçosa.

Professor at the Federal Institute of Northern Minas Gerais - Januária Campus.

E-mail: ddpacheco.agro@gmail.com

Danley Freitas Gonçalves

Bachelor's degree student in Agronomic Engineering at the Instituto Federal University of Northern Minas Gerais - Januária Campus.

E-mail: dannfreitas13@gmail.com

Rafael Tálison Magalhães Campos

Graduate of the Agronomic Engineering course at the Federal Institute of Northern Minas Gerais - Januária Campus.

e-mail: Rafael.talison5@gmail.com

Rafael Viana de Souza

Agronomist graduated from the Federal Institute of Northern Minas Gerais - IFNMG Januária Campus.

e-mail: rafinhaviana877@gmail.com

Hamilton dos Reis Sales

Master's degree in Biology and Conservation from the State University of Montes Claros. Graduated in Agronomic Engineering from the Federal Institute of Northern Minas Gerais - IFNMG Januária Campus.

e-mail: hamiltonbioflora@gmail.com

Ernesto Filipe Lopes

Agronomist graduated from the Federal Institute of Northern Minas Gerais - IFNMG Januária Campus.

e-mail: ernesto-lopes@live.com

ABSTRACT - Unlike techniques for raising soil pH, there are few agronomic alternatives for reducing the pH of alkalinised soils in order to make them productive. Irrigation with water rich in calcium carbonates alkalises soils in the Brazilian semi-arid region, as the pH rises above the ideal range for cultivation (5.7 to 6.5), drastically reducing the availability of nutrients such as P, Fe, Mn, Zn, S, B and Cu. The aim of this study was to assess the efficiency of doses of elemental sulphur (S^0) and cattle manure in correcting the pH of alkalinised soils in the north of Minas Gerais. The doses tested were 0; 0.5; 1.0; 2.0 and 4.0 t ha^{-1} of S^0 combined with 0, 10, 15, 25 and 50 t ha^{-1} of cattle manure using the Double Square experimental matrix, and the characteristics assessed were active acidity (pH in water), total acidity (H+Al), electrical conductivity, Ca^{2+}, Mg^{2+} at different dates. As a result, $S°$ lowered the soil pH, reaching a minimum value of 6.52. This was due to the release of H^+ ions into the soil solution after their oxidation to so_4^{2-}. The manure potentiated this acidification on the 30th day, possibly by releasing organic acids. The concentrations of H+Al increased as the doses of S^0 increased. The values of Ca^{2+} and Mg^{2+} decreased in response to the reduction in soil pH, indicating that they migrated in the profile beyond the range sampled. The soil's electrical conductivity increased in response to the treatments, possibly due to the increase in soluble ion concentrations following the conversion of S^0 to so_4^{2-}, and the saline impact of this on crops is something that deserves due attention. It was concluded that S^0 is a powerful soil pH reducer and should be evaluated in association with different organic matter sources and soil types for better efficiency and recommendations.

Keywords: active acidity, potential acidity, electrical conductivity, oxidation.

CHAPTER 1

INTRODUCTION

Although there are consolidated agronomic recommendations for raising soil pH, little information is available on reducing it (MAGALHÃES et al., 2005). Measures are therefore needed to reduce or minimise the alkalinity caused by excess carbonates in the soil and to allow for greater efficiency in the use of fertilisers and the availability of nutrients to plants in general.

The use of alkaline irrigation water rich in calcium carbonates and the presence of calcareous rocks in the soils of the Brazilian semi-arid region lead to the formation of unsuitable conditions for agricultural crops (RIBEIRO, M. S 2010), especially by raising the soil pH above the ideal range for cultivation (5.7 to 6.5). In alkaline soil pH conditions, there are severe changes in the chemical balance of nutrients in the soil solution (HASHEMIMAJD 2012).

For Brady (1989), the soil is alkaline, sometimes pronouncedly so, especially when there is sodium carbonate, and it is not uncommon for the pH to reach 9 or even 10. Alkaline soils are naturally characteristic of most arid and semi-arid regions. Alkalinity occurs when most of the pH-dependent negative charges are saturated by bases, which dislodge hydrogen, which passes into the soil solution. The reaction of the soil is important because it influences the availability of nutrients, providing favourable or toxic conditions. It also regulates the activity of microorganisms, which carry out useful transformations to improve soil conditions or, conversely, to cause plant diseases.

Soils with a pH below 5 tend to be deficient in Ca, Mg, P, Mo and B, and toxic to Al, Mn and heavy metals. On the other hand, a high pH, between 8.0 and 8.5, indicates the occurrence of calcium carbonate and/or free magnesium and low availability of the elements P, Fe, Mn, Zn, B, S and Cu.

Correcting alkaline soils involves washing with irrigation water and using products such as agricultural gypsum, sulphuric acid, ferrous sulphate, elemental sulphur and aluminium sulphate. One corrective that is widely used to reduce the pH of mineral soils in regions with alkalinity problems is elemental sulphur (S^0). Its acidifying effect is associated with its oxidation by microorganisms, with the consequent formation of sulphuric acid and subsequent release of hydrogen ions into the solution (HEYDARNEZHAD et al., 2012).

The microorganisms involved in the oxidation process are mainly bacteria of the genus Thiobacillus thiooxidans, first forming sulphuric acid which is converted to sulphate and the release of two H ions$^+$ (HEYDARNEZHAD et al., 2012). In addition to microorganisms, physical factors in the substrate influence the rate of S oxidation, such as humidity, aeration, temperature and organic matter content (ORMAN & KAPLAN, 2011). However, high humidity levels significantly reduce oxidation as a result of the substrate's low aeration.

Organic matter in the soil is continually being broken down by microorganisms into organic acids, carbon dioxide (CO_2) and water, forming carbonic acid (H_2CO_3). Carbonic acid in turn reacts with calcium and magnesium carbonates in the soil to form soluble bicarbonates which are leached out, leaving the soil more acidic. Higher levels favour oxidation, which is associated with its use as a source of energy and carbon by microorganisms (SIERRA et al., 2007).

There is little research into the use and doses of S^0 and organic fertiliser to regulate soil pH in order to make the environment more favourable for plant growth. The aim of this study was to assess the effect of S^0 and organic fertilisation with cattle manure on the pH of alkaline soils in the soil and climate conditions of northern Minas Gerais.

CHAPTER 2

LITERATURE REVIEW

2.1. SULPHUR AS A NUTRIENT

S was recognised as a plant nutrient over 200 years ago (DUKE and REISENAUER, 1986). Its deficiency is a limiting factor for agricultural production in large areas of Brazil, particularly in the Cerrado region.

The sources of S most commonly used in fertilisation are simple superphosphate (12% S-sulphate) and ammonium sulphate (24% S-sulphate), either on their own or as components of commercial formulas with low NPK concentrations. On the other hand, commercial formulas with high concentrations of NPK use raw materials such as urea, triple superphosphate and potassium chloride, which contain trace amounts of S, and this can induce S deficiency in plants.

2.2. FUNCTIONS OF SULPHUR IN PLANTS

S is absorbed by plants mainly in the form of SO_4^{-2} diluted in the soil solution, but it can also enter plants through the leaves in the form of sulphur dioxide (SO_2) dissolved in the air. The element is part of every living cell and is a constituent of amino acids that make up proteins. Other functions of S in plants are: to help form enzymes, proteins, vitamins and chlorophyll; to promote nodulation for nitrogen fixation by legumes; to compose volatile organic substances, responsible for the characteristic odour of garlic, mustard, onions and other vegetables; to accelerate the ripening and quality of seeds and fruit; to improve water use efficiency; and to control soil-borne diseases.

2.3. SULPHUR IN THE SOIL

S is the thirteenth most abundant element in the earth's crust (Bissani & Tedesco, 1988), with a content of around 0.06% (Barber, 1995). It occurs in solid compounds such as soluble and insoluble salts and in the form of gases. The main minerals containing this element in rocks and soils are gypsum, epsomite, mirabilite and pyrite (Tisdale et al., 1993). The weathering of these, through physical, chemical and biological processes, releases S to microorganisms and plants. Rainwater (especially in areas close to industrial centres), fertilisers and pesticides can also be sources of the element.

In the soil's source material, silicate minerals contain an average of 0.01% S, but this content can be higher in biotite, chlorite and other minerals. In igneous rocks, the S content ranges from 0.02 to 0.07%, while in sedimentary rocks it ranges from 0.02 to 0.22% (Tisdale et al., 1993).

S occurs in the soil in organic and inorganic forms (Freney, 1986). The proportion between these two forms of S in the soil varies with the type of soil and its depth. In the surface horizons of soils, especially in tropical environments, organic S makes up most of the total S. However, with increasing depth, organic S makes up the majority of the total S in the soil. However, with increasing depth, organic S decreases as organic matter decreases (Duke & Reisenauer, 1986). In surface horizons of various Brazilian soils, Neptune et al. (1975) found that S-organic constituted 77 to 95 per cent of total S.

2.4.   OXIDATION OF ELEMENTAL S

S^0 is the pure form of the element, with at least 98 % S. In nature it occurs mainly in volcanic deposits, sediments, coal, oil, natural gas, in the form of organic compounds and is a by-product recovered in oil and natural gas refineries. The oxidation of S^0 is mainly carried out by soil microorganisms, whose activity is regulated by temperature, texture, aeration, pH and soil fertility (organic matter and nutrients). Another factor that affects the oxidation of S^0 is its specific surface area. This depends on the shape, size, composition, degree of dispersion and application rate of the S-elemental particles. The greater the specific surface area, associated with the presence of oxidising microorganisms, the greater the oxidation rate, transforming S^0 into S-sulphate (Germida & Janzen, 1993).

The fertilising action of S^0 depends on its rate of oxidation to S-sulphate (Janzen & Bettany, 1987a). Saik (1995) described the oxidation reaction of S^0 in the soil as follows:

$$\text{S-elementar} + 1\tfrac{1}{2}O_2 + H_2O \rightarrow 2H^+ + SO_4^{2-}$$

The intensity of this reaction is related to biological, physical and chemical factors in the soil, as well as the properties of the fertiliser containing the S^0 (Germida

& Janzen, 1993).

2.5.   MICRORGANISMS

The oxidation of S^0 is mainly carried out by microorganisms, although this reaction can occur due to abiotic factors (Germida & Janzen, 1993). The biological oxidation of S^0 must follow the order of the products formed (Nor & Tabatabai, 1977):

$$S^0 \rightarrow S_2O_3^{-2} \rightarrow SO_2 \rightarrow SO_4^{-2}$$

For biological S^0 oxidation to occur, the specificity of the microorganisms and the size of their population are important. Research carried out in Canada on 28 soils showed a positive correlation between the oxidation of S^0 with microbial biomass and also with the rate of soil respiration. Therefore, the size of the specific microbial population and its activity determine the rate of S^0 oxidation (Lawrence & Germida, 1988). Furthermore, sulphate oxidation, being dependent on microorganisms, results in a system in which favourable conditions for plant growth are generally also favourable for microbiological oxidation (Watkinson & Blair, 1993).

Various groups of microorganisms in the soil can oxidise S^0 . These can be divided into the following groups: a) chemoautotrophs, such as bacteria of the genus Thiobacillus; b) photoautotrophs; c) heterotrophs (bacteria and fungi). In most aerobic soils, organisms in groups a and c are the most important (Germida & Janzen, 1993).

The peculiarities of the rhizosphere of plant species influence the oxidation rate of S^0 , but little research has been done on the subject (Germida & Janzen, 1993). Grayston & Germida (1990) observed an increase in the amount of S^0 oxidising microorganisms in the rhizosphere of plants. Janzen (1990) concluded that the cultivation of different plant species had practically no effect on the rate of S oxidation^0 compared to fallow soils. These contradictions can be partially explained by: 1) the rhizosphere may have little influence on the oxidation of S^0 in relation to the rest of the soil volume; 2) a higher population of S^0 oxidising microorganisms observed by Grayston & Germida (1990) may not lead to a higher oxidation rate (Janzen, 1990).

2.5.1. The microbiology of S-element oxidation

Several chemoautotrophic bacteria of the genus Thiobacillus oxidise reduced S compounds. Among these bacteria, some species oxidise S^0 : T. thiooxidans, T. ferrooxidans, T. neapolitanus, T. kabobis, T. denitrificans, T. perometabolis and T. thioparus ((Wainwright, 1984; Germida & Janzen, 1993). They differ in their physiological characteristics and use reduced compounds as a substrate for growth. Most of these species are obligate aerobes, with the exception of T. denitrificans (Germida & Janzen, 1993). The oxidising capacity of the different species of bacteria is related to the pH of the medium. The species Thiobacillus thioparus and T. neapolitanus oxidise S^0 when the pH is close to neutral, while T. thiooxidans predominates at pHs below 5 and can grow at very acidic pHs around 2 (Wainwright, 1984).

The genus Thiobacillus is considered to be the most important in the oxidation of S^0 in soils. However, there are few studies on the subject. This genus of bacteria is easily isolated in specific and extreme environments, such as acidic, hot and polluted soils, close to waste with a high concentration of S^0 (Wainwright & Killham, 1982). However, attempts to isolate these microorganisms in agricultural soils, in which oxidation of S^0 occurs, have been unsuccessful. In other soils, where the addition of S^0 increases the population of Thiobacilli, this growth was not an adequate indicator for the rate of S^0 oxidation. This casts doubt on whether the Thiobacilli are the main S^0 oxidising microorganisms in the soil (Germida & Janzen, 1993).

The oxidation of S^0 is mediated by various heterotrophic microorganisms (Wainwright & Killham, 1980). These include the bacteria Achromobacter sp., Bacillus brevis and Micrococus spp; the actinomycetes Spretomyces spp; and the fungi Absidia glauca, Fusarium solani and Penicillium decumbens. There is evidence that in agricultural soils the majority of S-oxidising microorganisms[0] are heterotrophic, but it is difficult to recognise the function of these different groups (Germida & Janzen, 1993).

2.5.2. Temperature

The microbiological oxidation rate of S^0 is affected by temperature. At the temperatures normally observed at the soil surface, the relationship between oxidation rate and soil temperature is exponential. When compared to other soil factors that influence the oxidation rate of S^0 , temperature is the most important factor (GERMIDA and JANZEN, 1993). Although the optimum temperature for oxidation is not yet well defined, it may vary depending on the characteristics of the soil. Skiba and Wainwright (1984) suggest that the highest oxidation rates occur between 30 and 40°C. Oxidation is nil or very slow at temperatures below 5°C (NOR and TABATABAI, 1977; SKIBA and WAINWRIGHT, 1984; WAINWRIGHT, 1984; JANZEN and BETTANY, 1987c).

The management of residues applied to the soil, soil colour, depth and time of fertiliser application all influence the oxidation rate of S^0 . In addition, the high temperature sensitivity of the microorganisms that oxidise S^0 indicates that the efficiency of fertilisers containing S^0 varies greatly between climatic regions (GERMIDA and JANZEN, 1993). Therefore, in cold regions it is possible that the efficiency of fertilisers containing S^0 is low (JANZEN and BETTANY, 1987c).

2.6.   SOIL PHYSICAL PROPERTIES

S oxidation is affected by soil physical properties, although these effects are not always consistent (GERMIDA and JANZEN, 1993). In some investigations, S oxidation rates^0 were inversely related to clay content and directly to sand content (JANZEN and BETTANY, 1987a; LAWRENCE and GERMIDA, 1988; DENG and DICK, 1990). Other researchers have found no significant relationship between soil texture and the rate of S oxidation (RHEM and CADWELL, 1968; McCASKILL and BLAIR, 1987; WATKINSON, 1989). However, poorly structured soils tend to have a lower S oxidation rate^0 when compared to better structured soils (LEE et al., 1987; WATKINSON, 1989). Many of these observations are probably indirect reflections of the state of soil aeration (GERMIDA and JANZEN, 1993), which affects the microbiological oxidation reactions of S .^{0}

2.6.1. Humidity and aeration

The oxidation rate of S^0 is related to soil water potential, showing a parabolic relationship up to the maximum rates (JANZEN and BETTANY, 1987c). Maximum rates occur around field capacity and decrease at higher or lower potentials. This model reflects the interactive effects between the availability of water and oxygen in the soil (BURNS, 1967). In conditions of low soil moisture, oxidation is limited by insufficient water for microbial activity. Conversely, in soils with high moisture content, oxidation of S^0 is limited by inadequate aeration. Therefore, in soils cultivated with flooded rice, for example, S^0 should not be used, as oxidation occurs very slowly due to the low oxygen content in the soil.

2.6.2. Soil pH

The pH is often positively related to the rate of S oxidation^0 (NOR and TABATABAI, 1977; JANZEN and BETTANY, 1987b; LAWRENCE and GERMIDA, 1988). At a higher pH, the soil buffers the sulphuric acid formed during oxidation and so the activity of the microorganisms that transform S into so_4^{2-} is not compromised (FOX et al., 1964; BARROW, 1971). For a dystrophic red latosol typical of the Cerrados region, Horowitz (2003) showed that there is a positive relationship between the oxidation of S^0 and soil pH.

2.6.3. Organic matter content and nutrient availability

Several studies have shown a positive effect between the oxidation of S^0 and the organic matter content of the soil. This can be attributed to the response of heterotrophic organisms that oxidise S^0 using the available substrate as an energy source (WAINWRIGHT et al., 1986; LAWRENCE and GERMIDA, 1988; CIFUENTES and LINDEMANN, 1993; COWELL and SHOENAU, 1995).

Studies have shown that adequate P availability in the soil stimulates S^0 oxidation (JANZEN and BETTANY, 1987a; LAWRENCE and GERMIDA, 1988). The absence of a well-defined relationship between the oxidation of S^0 and the various physical and chemical properties of the soil does not necessarily imply that these effects are not important. It is likely that oxidation is governed by the interaction and integration of these factors and not just by one factor in isolation, except in extreme cases (GERMIDA and JAZEN, 1993). In general, it is observed that the soil factors that have a stimulating effect on S^0 oxidation coincide with those that promote good plant development.

2.6.4. S particle size, shape and distribution[0]

Smaller particles of S^0 favour higher oxidation rates in the soil, which is explained by the fact that the increased surface area of the particles stimulates contact and the activity of oxidising microorganisms (WAINWRIGHT, 1984). In general, for rapid oxidation of S^0, the particles of this fertiliser must be less than 0.15 mm in size.

The shape of the S particle also affects its specific surface area (Watkinson, 1989). As spheres have a lower surface area per mass ratio, any departure from sphericity will increase the specific surface area (McCaskill & Blair, 1987). For example, particles of S^0 in the shape of blades will have a greater specific surface area than particles in the shape of spheres or in the shape of blocks (Germida & Janzen, 1993). Therefore, sphere-shaped particles will have a more gradual reduction in surface area than blade-shaped particles (Watkinson, 1989).

The distribution of S^0 in the soil can reduce oxidation due to its hydrophobic character. Thus, in grouped particles, hydration may be deficient for oxidation to occur (JANZEN and BETANNY, 1986; WATKINSON, 1989). With greater dispersion of the S^0 particles in the soil, the oxidation rate increases. The slow oxidation of poorly dispersed particles in the soil is of significant importance for the efficiency of fertilisers containing S. Several studies have shown that the highest oxidation rates of a fertiliser containing S^0 occur when it is mixed with the soil, rather than banded (CHIEN et al., 1988).

2.7. PH AND NUTRIENT AVAILABILITY

When the soil pH rises from 4 to 5, there is a 10-fold reduction in the concentration of H^+ in the soil solution. The decrease in H^+ is in the order of 100 times when the pH rises from 4 to 6. As a result of the activity of H^+, the pH of the soil drastically affects the availability of nutrients and, in general, the pH should be between 6.0 and 6.5 in order to achieve the right balance between nutrients. Soils with a pH between 5.8 and 7.5 tend to be problem-free from the point of view of plant growth. Below pH 5, there may be deficiencies of Ca, Mg, P, Mo and B; toxicity of Al, Mn, Zn and other heavy metals. The presence of pH between 8.0 and 8.5 indicates the occurrence of free calcium and/or magnesium carbonate and low availability of the elements P, Fe, Mn, Zn and Cu.

CHAPTER 3

MATERIAL AND METHODS

The experiment was conducted in the municipality of Januária, on the campus of the Federal Institute of Northern Minas Gerais Januária Campus. The municipality is located at **15°29' south latitude, 44°21' west longitude and 434 metres above sea level.**

The soil in the experimental area is a Yellow Red Latosol, whose physical-chemical characterisation of samples collected in the 0-20 cm depth layer is shown in Table 1. Inside, the experimental area was cultivated with different genomic groups of banana trees in a micro-sprinkler irrigated system.

Table 1 - Physical and chemical characterisation of the soil sample in the 020 cm depth layer before the experiment was set up. IFNMG, Januária Campus, 2016.

PH	MO	P	K	Ca	Mg	Al	H+Al	Ass	Fe	Mn	Zn	Prem	EC	Sand	Silt	Clay
	dag/kg	..mg/dm³	cmolc/dm³............					.............mg/dm³........				mg/l	dS/m	..........dag/kg........		
8,02	0,3	26,6	50	2,9	0,2	0	0,80	0,5	13,3	29,5	2	38,01	0,2	74	7	19

Extractors: pH: H_2O (1:2.5); P, K, Zn, Mn, Fe and Cu: Mehlich-1x; Ca^{2+}, Mg^{2+} and Al^{3+} : KCl 1 mol L^1 . H+Al: pHSMP ; M.O (Organic Matter): colorimetry; CE (electrical conductivity); textural analysis: beaker method.

The treatments were defined by the factors S^0 and cattle manure in combined doses according to the Double Square experimental matrix (ALVAREZ, 1985). The doses tested were 0; 0.5; 1, 2 and 4 t ha⁻¹ of S^0 and 0, 10, 15, 25 and 50 t ha⁻¹ of manure, totalling 13 treatments in a randomised block design with three replications. The experimental units were 1 m wide and 1 m long and their distribution in the area followed the sketch shown in Figure 1.

The doses of S^0 and manure, according to the treatment, were applied to the entire area of the experimental unit and incorporated superficially into the soil using a hoe.

Weeding was carried out around the experimental unit whenever necessary. Irrigation was carried out daily in the morning for 3 hours with a blade of approximately 3.5 mm, in order to meet the water requirements of the banana crop already established in the area.

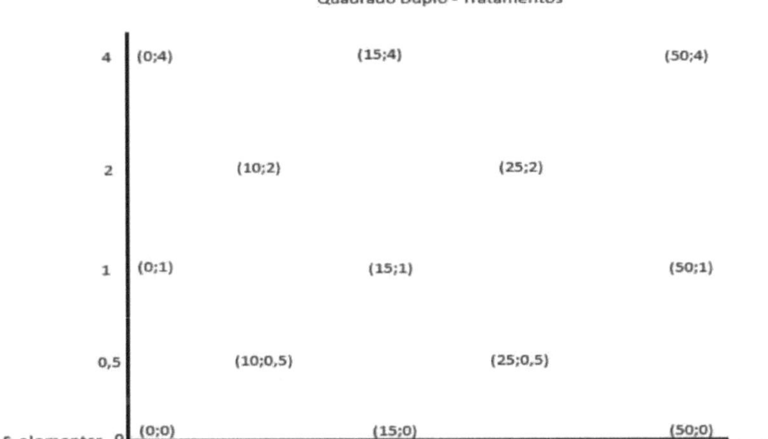

Figure 1: Treatments defined by the combination of doses of S^0 and cattle manure following the Double Square experimental matrix. IFNMG, Januária campus, 2016.

The experimental plots were sampled at 30, 60, 90, 184, 214, 244 and 274 days after the treatments were applied. For this purpose, composite samples were collected inside the experimental units, each consisting of a mixture of two simple deformed soil samples in the 0 to 20 cm layer using a Dutch auger, always at 1.30pm.

The soil samples were sent to the soil laboratory at IFNMG, Januária, where they were dried in the shade, sieved through a 2 mm sieve and assessed for active acidity (pH in water), total acidity (SMP pH and H+Al), electrical conductivity, calcium and magnesium following the methodological recommendations of the SOIL FERTILITY COMMISSION OF THE STATE OF MINAS GERAIS (1999).

The data on pH in water, SMP pH, H+Al, calcium, magnesium and electrical conductivity were analysed for variance in order to quantify the precision of the experiment. The data was then analysed by regression, considering them as dependent variables of the doses of S^0 and manure, generating response surfaces. The models were selected on the basis of the highest significance of the regression parameter coefficients, using SAEG software.

CHAPTER 4

RESULTS AND DISCUSSION

The summary of the analysis of variance for the characteristics of pH in water, SMP pH, H+Al, calcium, magnesium and electrical conductivity showed a significant effect for treatments and blocks and, in general, the coefficients of variation were below 20%, indicating adequate precision in the field experiments (Tables 2, 3, 4 and 5). The significant effect for treatments indicates that the characteristics evaluated were influenced by the S and manure factors. The significant effect for blocks indicates that the experimental area was heterogeneous and in this condition local control was essential to allow for acceptable experimental error.

Table 2 - Summary of analysis of variance for data on pH in water, SMP pH, H+Al and electrical conductivity (EC) in soil samples collected in the 0 to 20 cm depth layer at 30 days after application of the treatments. IFNMG, Januária campus, 2016.

Sources of Variation	gi	Mean Square			
		pH Water	SMP pH	H+Al	EC
Treatment	12	0,45768**	0,03292***	0,04664***	0,03414***
Block	2	0,34237**	0,03208**	0,04293**	0,00671^{ns}
Waste	24	0,04654	0,00543	0,00910	0,00449
CV(%)		2,83	1,01	10,08	24,80

ns, not significant; *,**, *** significant at 5%, 1% and 0.1% probability levels.

Table 3 - Summary of analysis of variance for data on pH in Water, SMP pH, H+Al and electrical conductivity (EC) in soil samples collected in the 0 to 20 cm depth layer at 60 days after application of the treatments. IFNMG, Januária campus, 2016.

Sources of Variation	gl	Mean Square			
		pH water	pH SMP	H+Al	EC
Treatment	12	0,33084***	0,05441***	0,09560**	0,12190***
Block	2	0,02407^{ns}	0,03730**	0,05523**	0,03100^{ns}
Waste	24	0,06177	0,00721	0,01306	0,01927
CV(%)		3,39	1,17	11,00	43,47

ns, not significant; *,**, *** significant at 5%, 1% and 0.1% probability levels.

Table 4 - Summary of analysis of variance for data on pH in Water, SMP pH, H+Al and electrical conductivity (EC) in soil samples collected in the 0 to 20 cm depth layer 90 days after application of the treatments. IFNMG, Januária campus, 2016.

Sources of Variation	gl	Mean Square			
		pH water	SMP pH	H + Al	EC
Treatment	12	0,79323***	0,11756**	0,21419** 1,48489**	0,23928***
Block	2	0,06370^{ns}	0,91911***	*	0,01793^{ns}
Waste	24	0,14993	0,04333	0,08988	0,01618
CV(%)		5,33	2,88	27,02	29,21

ns, not significant; *,**, *** significant at 5%, 1% and 0.1% probability levels.

Table 5 - Summary of analysis of variance for data on pH in water, H + Al, Ca, Mg and electrical conductivity (EC) in soil samples collected in the 0 to 20 cm depth layer at 184, 214, 244 and 274 days after application of the treatments. IFNMG, Januária campus, 2017.

FV	Gl	Mean Square				
		pH H2O	H+Al	Ca	Mg	EC
Treatment	12	3,1292***	0,1346*	1,5808**	0,0134**	2,093***
Block	2	1,558***	0,2727*	0	0	0
Waste	141	0,1785	0,0671	0,1737	0,0026	0,0705
CV (%)		5,657	27,663	9,874	15,97	22,545

ns, not significant; *,**, *** significant by F test at 5; 1 and 0.1% probability levels

The pH of the soil, 30 days after the application of the treatments, decreased in response to the doses of S and manure (Figure 3). The highest value was 8.16, in the absence of S and manure applications; and the lowest, 6.93, was estimated with the combined doses of 4 t ha^{-1} S and 30 t ha^{-1} of manure. Therefore, S and manure enabled an adequate soil acidification response, and this was a rapid process. The sudden increase of more than 10 times in the concentration of H^{+} in the soil solution with the addition of S^{0} and manure is important in order to increase the availability of nutrients, especially the cationic micronutrients Cu, Fe, Mn and Zn. The availability of cationic micronutrients tends to be extremely low in alkaline conditions due to the high concentration of OH^- . These react with the aforementioned elements precipitating them as $Cu(OH)_2$, $Fe(OH)_3$, $Mn(OH)_4$ and $Zn(OH)_2$, $Cu(OH)_2$ (International Handbook of Soil Fertility, 1998). Thus, the soluble concentrations of Cu^{2+} , Fe^{2+} , Mn^{2+} and Zn^{2+} decrease drastically in the soil, inducing deficiencies. Lawrence and Germida (1988) indicate that the microbial decomposition of organic matter enhances the oxidation of S^0 , which increases acidity, preventing the precipitation of these cationic micronutrients. Thus, the authors conclude, similarly to the present study, that a reduction in soil pH can be achieved by adding organic matter and S^0 to the medium.

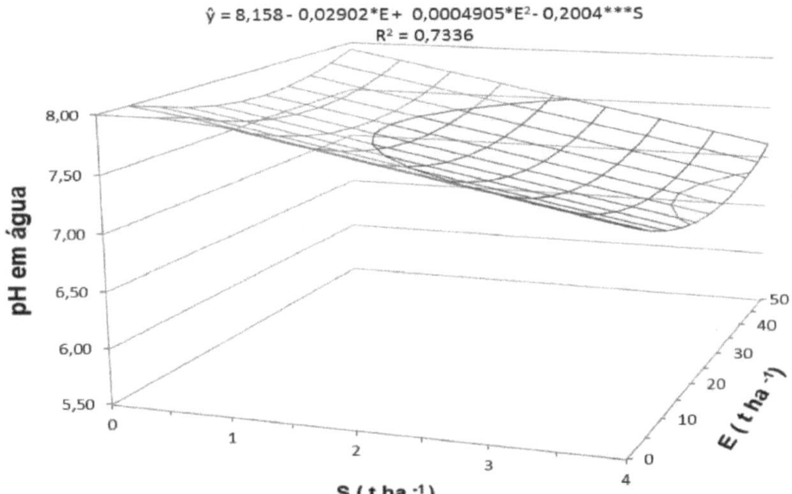

Figura 3 - Response surface for pH data in water at 30 days in response to S doses and bovine manure (IFNMG, Januária campus, 2016).

At 60 days after the treatments were applied, the pH in water decreased in response to the doses of S^0 , and was not influenced by the doses of manure (Figure 4). The highest value was 7.69, in the absence of S and manure applications; while the lowest, 7.00, was estimated with the 4 t ha⁻¹ dose of S, regardless of the manure dose. The pH value, under the desired acidity conditions, was very close in the first two seasons of soil sampling, and this is important because it shows that the buffer capacity

19

of the soil was unable to re-establish the alkalinity of the medium in the interval between the 30th and 60th day of the experiment. Thus, the S source[0] , under the conditions of the best doses, favours acidifying conditions in the soil, maximising the availability of most nutrients.

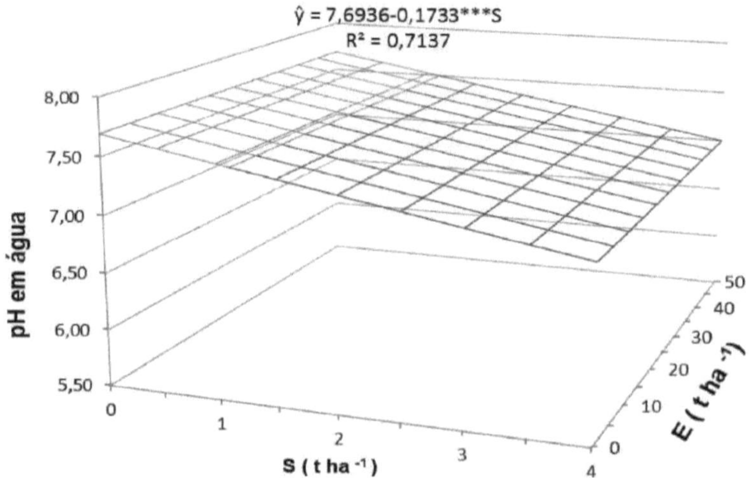

Figura 4 - Response surface for pH data in water at 60 days in response to S doses and bovine manure (IFNMG, Januária campus, 2016).

The active acidity of the soil assessed at 90 days showed a significant response only to the doses of S^0 , similar to the assessment at 60 days (Figure 5). The highest pH value was 7.78, in the absence of S^0 and manure applications; and the lowest, 6.57, was estimated with the application of 4 t ha^{-1} S^0 , regardless of the manure dose. This evaluation showed the lowest soil pH over the first three analyses, with 6.57 being the most suitable pH as a soil reaction to balance nutrient availability. This result, combined with the analyses at 30 and 60 days, shows that the acidification of the soil is relatively rapid, as there had already been a sharp decrease in pH at just 30 days, and that it is continuous, extending up to the 90th day of application of the treatments.

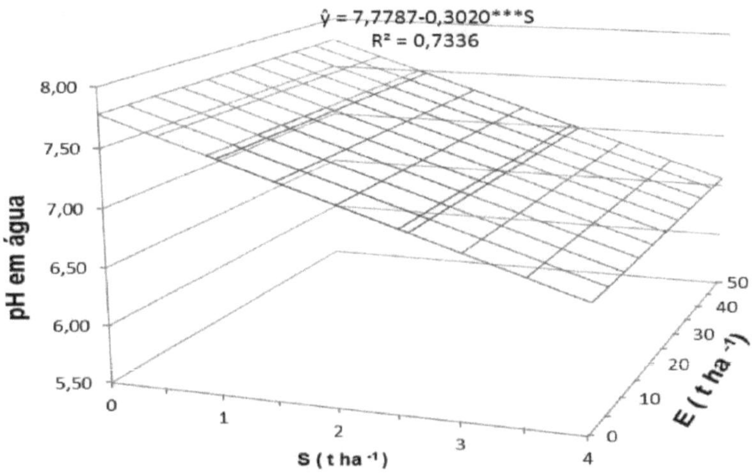

Figura 5 - **Response surface for water pH data at 90 days in response to S doses and bovine manure (IFNMG, Januária campus, 2016).**

The soil pH values in samples taken at 184 and 214 days decreased in response to the doses of S^0 , but were not influenced by the doses of manure (Figures 6 and 7). The highest pH value was 7.9, in the absence of S^0 and manure applications; and the lowest, 6.72, was estimated with the application of 4 t ha⁻¹ of S^0 in the absence of cattle manure (Figure 6). At 214 days, the maximum pH value was 8.25 and the minimum 6.99, respectively with applications of 0 and 4 t ha⁻¹ of S^0 (Figure 7).

$$\hat{y} = 7{,}8907 - 0{,}2923{***}X \quad R^2 = 0{,}9923$$

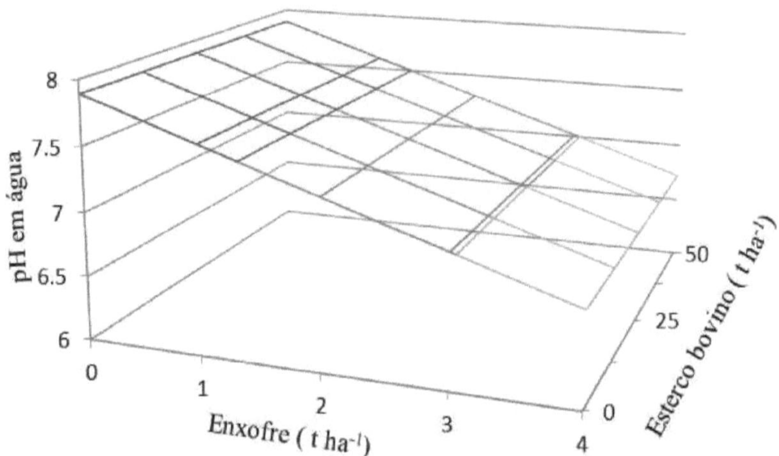

Figura 6 **Response surface for pH data in water at 184 days in response to doses of S[0] and bovine manure (IFNMG, Januária campus, 2016).**

$$\hat{y} = 8{,}2501 - 0{,}3143{**}X \quad R^2 = 0{,}9242$$

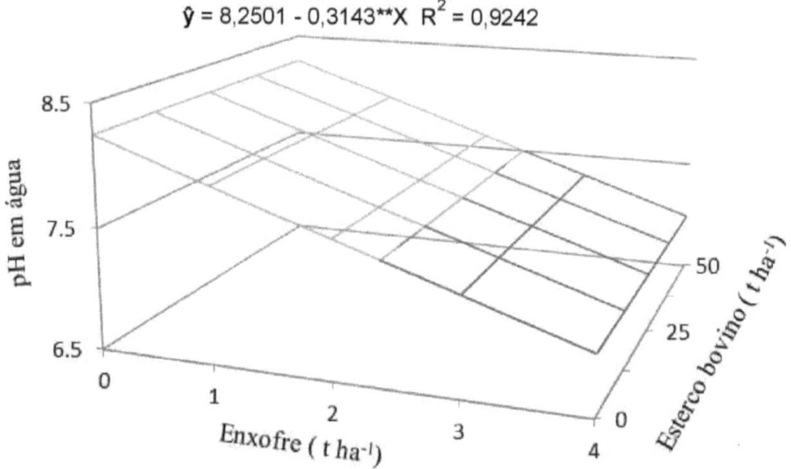

Figura 7 **Response surface for pH data in water at 214 days in response to doses of S[0] and bovine manure (IFNMG, Januária campus, 2016).**

The pH value in water for soil samples collected 244 days after application of the treatments decreased in response to the S^0 doses, and was not influenced by the manure doses (Figure 8). The highest value was 7.92, in the absence of S^0 and manure applications; while the lowest value, 6.52, was estimated with 4 t ha 0^{-1} s.

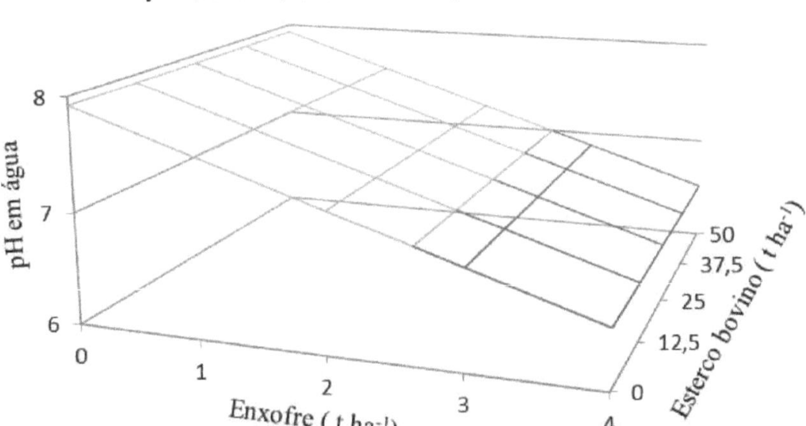

Figura 8   - Response surface for pH data in water at 244 days in response to doses of S⁰ and bovine manure (IFNMG, Januária campus, 2016).

The active acidity of the soil assessed at 274 days maintained the same behaviour as the assessment at 184 and 214 days (Figure 9). The highest pH value was 7.72, in the absence of S^0 and manure applications; and the lowest, 6.65, was estimated with the application of 4 t ha⁻¹ S^0 , regardless of the manure dose. The pH values in water were similar in the three soil sampling periods mentioned above, and this is important because it shows that the natural alkalinity of the environment is not re-established in the interval between 30 and 274 days of conducting the experiment. Thus, the S source⁰ , under the conditions of the best doses, provided acidifying conditions in the soil, and this is important in order to maximise the availability of most nutrients. The lowest soil pH was observed from 184 days to 274 days of analyses, with 6.52 being the most suitable pH as a soil reaction to balance nutrient availability (Figure 9).

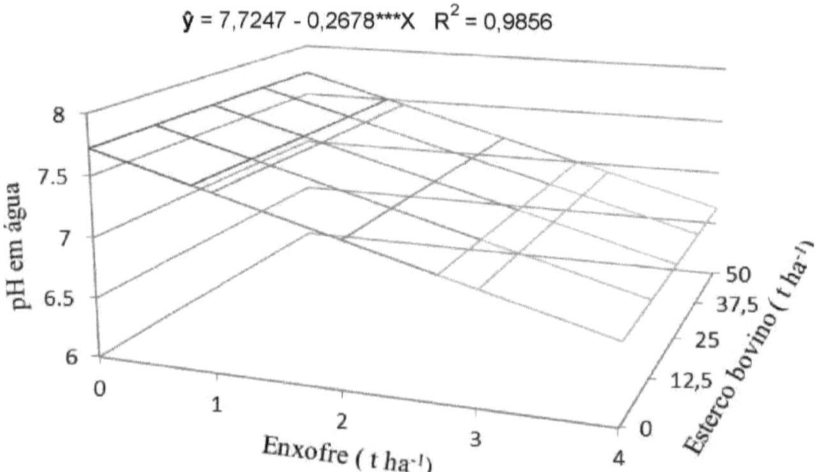

Figura 9 - **Response surface for pH data in water at 274 days in response to doses of S^0 and bovine manure (IFNMG, Januária campus, 2016).**

The reduction in pH caused by the action of S^0 is due to the biological oxidation of this product, which generates sulphuric acid in the soil (Stamford et al. 2007). The use of cattle manure also proved to be a potential soil pH reducer after 30 days, due to its decomposition by microorganisms into organic acids, carbon dioxide and water, forming carbonic acid. Carbonic acid in turn reacts with Ca and Mg carbonates in the soil to form soluble bicarbonates which are leached out and make the soil more acidic.

A positive relationship between S^0 oxidation and organic matter decomposition is common, and this can be attributed to the response of heterotrophic organisms that oxidise S^0 using the available organic substrate as an energy source (Wainwright et al., 1986; Lawrence & Germida 1988; Cifuentes & Lindemann, 1993; Cowell & Schoenau, 1995).

The SMP pH of the soil, 30 days after the application of the treatments, decreased in response to the doses of S^0 and manure (Figure 10). The highest value was 7.45 in the absence of S^0 and manure applications, while the lowest was 6.92 with the combined doses of 4 t ha^{-1} S^0 and 45 t ha^{-1} manure. Therefore, the joint action of S^0 and manure was an important means of ensuring soil acidification.

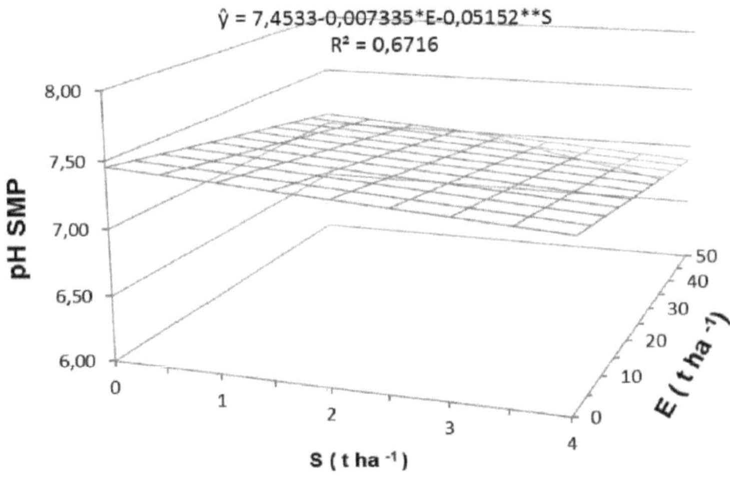

Figura 10 - **Response surface for SMP pH data at 30 days in response to S doses and cattle manure (IFNMG, Januária campus, 2016).**

In the second analysis, 60 days after the treatments were applied, the SMP pH decreased in response to the doses of S^0, and was not influenced by the doses of manure (Figure 11). The highest value was 7.37 in the absence of S^0 and manure applications, while the lowest 7.11 was estimated with the dose of 4 t ha^{-1} S regardless of the dose of manure applied.

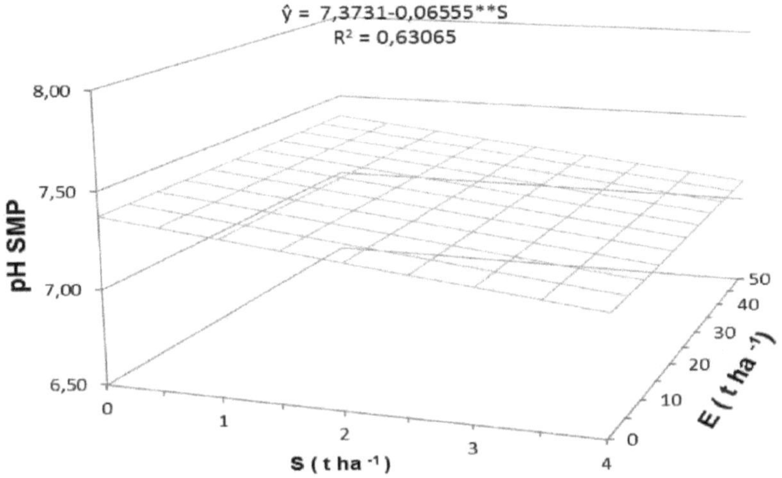

Figura 11 - **Response surface for SMP pH data at 60 days in response to S doses and cattle manure (IFNMG, Januária campus, 2016).**

At 90 days after the treatments were applied, the SMP pH decreased in response to the S doses and was not influenced by the manure doses, behaving in the same way as at 60 days (Figure 12). The highest value was 7.45 in the absence of S^0 and manure applications, while the lowest value was 6.95 with a dose of 4 t ha^{-1} S^0 regardless of the dose of manure applied.

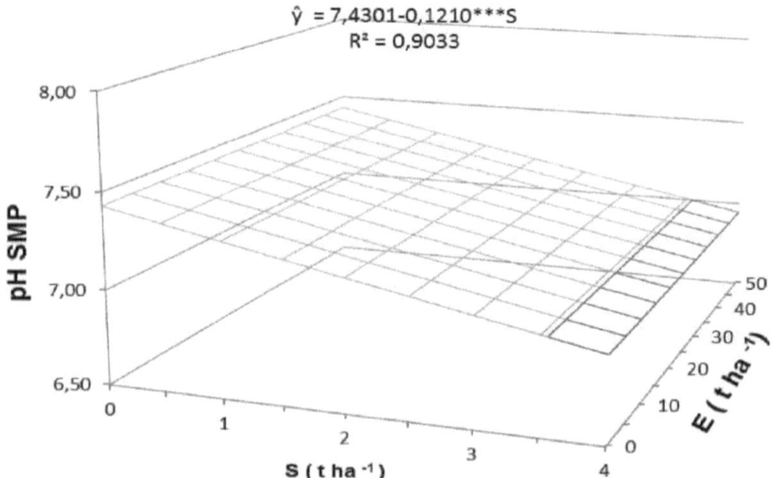

Figura 12 - **Response surface for SMP pH data at 90 days in response to S doses and cattle manure (IFNMG, Januária campus, 2016).**

The concentrations of H+Al increased as the doses of S increased[0] at 30 days after the treatments were applied (Figure 13). The lowest value was 0.79 cmolc dm^3 of H+Al in the absence of S and manure applications, while the highest was 1.03 cmolc dm3 at a dose of 4 t ha^{-1} S^0 , regardless of the dose of manure applied. As the soil had 0.8 cmolc.dm3 of H+Al in the analysis before the experiment was set up, it can be concluded that the H^+ concentrations increased in the soil, which is explained by the release of acidity into the soil solution after the oxidation of [0]

26

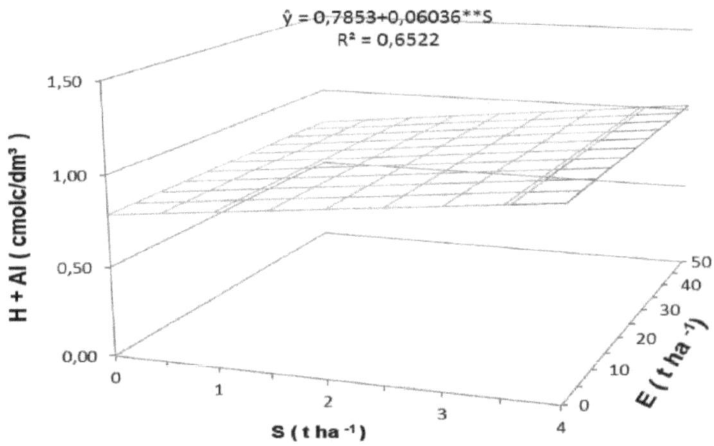

Figura 13   - Response surface for H+Al data at 30 days in response to S doses and cattle manure (IFNMG, Januária campus, 2016).

Subsequent analyses at 60 and 90 days showed a gradual increase in H+Al concentrations (Figures 14 and 15). H+Al practically doubled, indicating that the reduction in pH was caused by the change in H ions$^+$ in the soil solution. The element that increased acidity was S^0 , as at no stage of the evaluation was manure significant in raising H+Al concentrations.

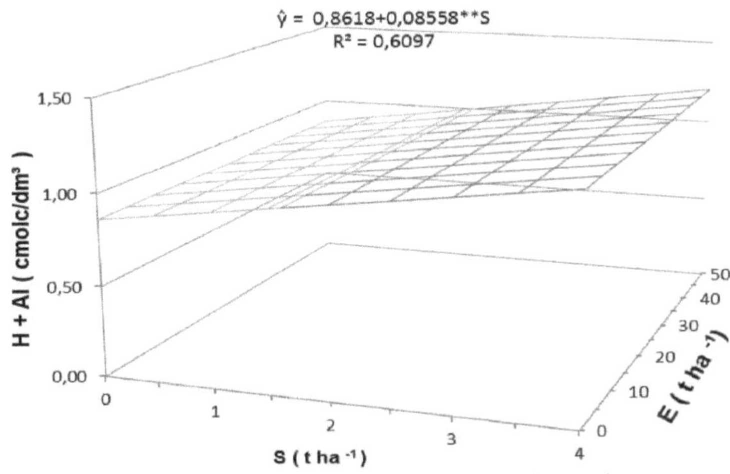

Figura 14   - Response surface for H+Al data at 60 days in response to S doses and cattle manure (IFNMG, Januária campus, 2016).

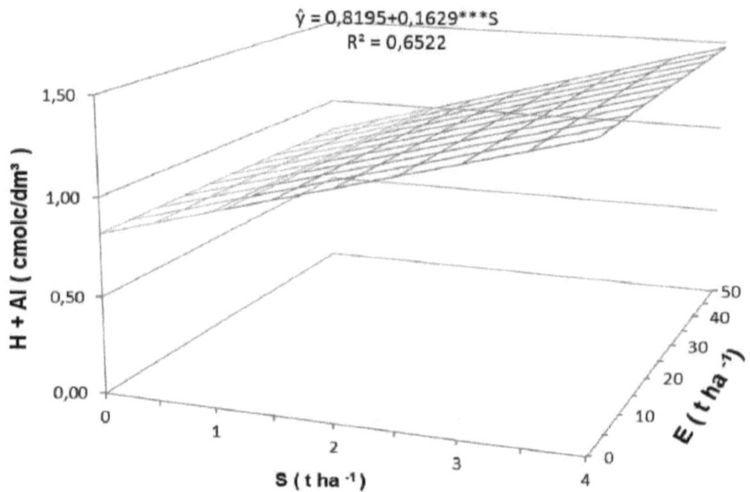

Figura 15   - Response surface for H+Al data at 90 days in response to S doses and cattle manure (IFNMG, Januária campus, 2016).

At 184 days, H+Al concentrations were close to 1.1 cmol.dm^{-3} with 4 t/ha of S, and their minimum was around 8.9 cmol.dm^{-3} in the absence of sulphur.

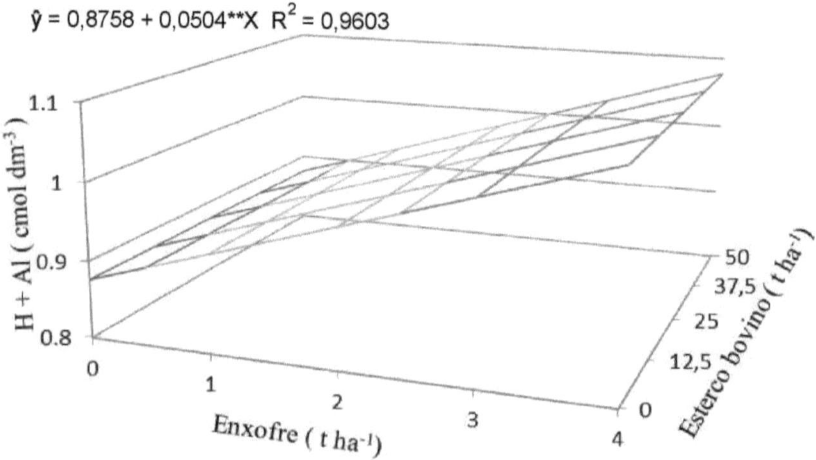

Figura 16   - Response surface for H+Al data at 184 days in response to doses of S⁰ and cattle manure (IFNMG, Januária campus, 2016).

At 214 days, the H+Al concentration stagnated in relation to the doses of sulphur and cattle manure (Figure 17), remaining at 1.15 cmol.dm .$^{-3}$

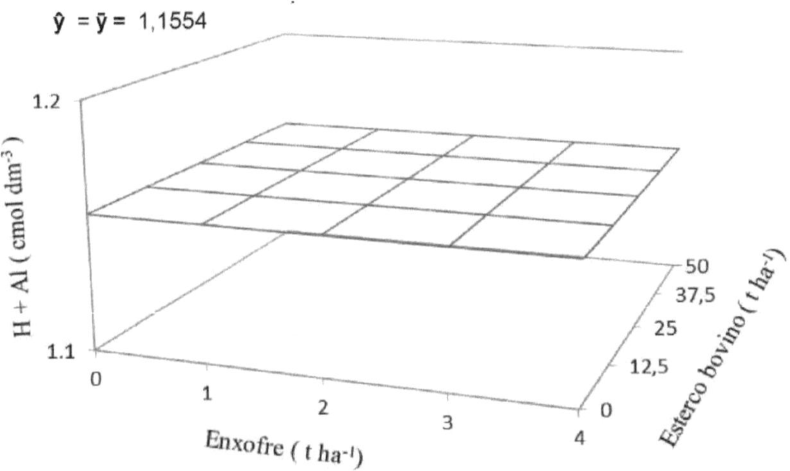

Figura 17 - Response surface for H+Al data at 214 days in response to doses of S^0 and cattle manure (IFNMG, Januária campus, 2016).

At 244 days, H+Al concentrations were affected by sulphur and cattle manure, with the lowest concentration found being 0.8 cmol.dm-3 of H+Al in the absence of sulphur and with 25 t/ha of cattle manure, and the highest being 1.2 cmol.dm-3 with 4 t/ha of sulphur and the absence or 50 t/ha of cattle manure (figure 18).

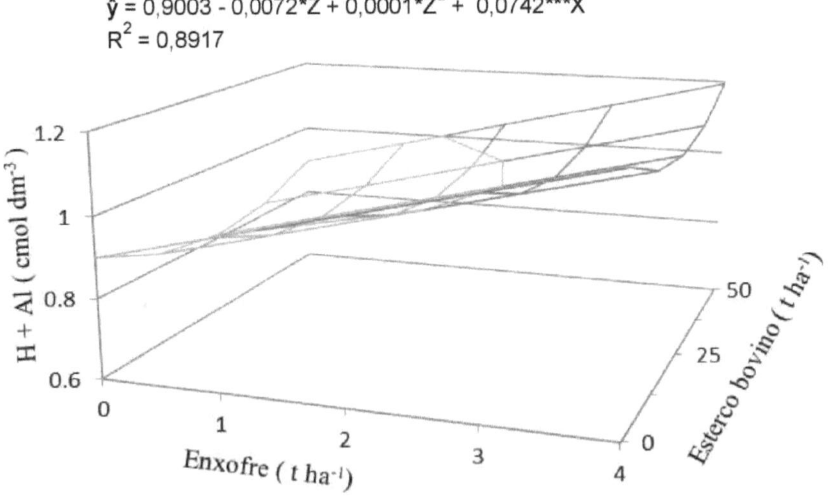

Figura 18   - Response surface for H+Al data at 244 days in response to doses of S⁰ and cattle manure (IFNMG, Januária campus, 2016).

The response of H+Al to the different doses of sulphur and cattle manure at 274 days shows an increase in concentrations in relation to the amount of sulphur, with a minimum of 0.5 cmol.dm-3 with 0 t/ha, and a maximum of approximately 0.9 cmol.dm-3 with 4 t/ha of sulphur.

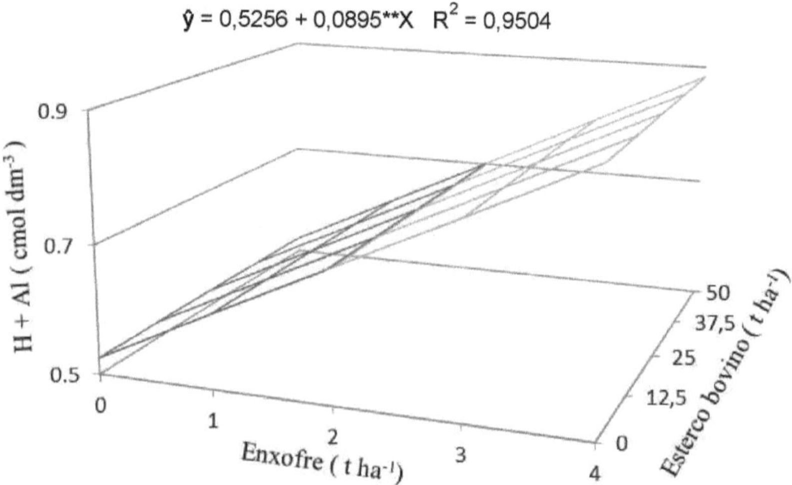

Figura 19   - Response surface for H+Al data at 274 days in response to doses of S⁰ and cattle manure (IFNMG, Januária campus, 2016).

The soil's electrical conductivity, which estimates the accumulated ionic concentrations, increased throughout the analysis period, especially in response to the doses of S⁰ (Figures 20, 21 and 22). In the first analysis, at 30 days, the soil EC almost doubled, indicating an increase in the concentration of salts in the soil solution (Figure 20). The values were 0.08 dSm⁻¹ in the absence of S and manure applications, and the lowest 0.44 dSm⁻¹ was estimated with the combined doses of 3 t.ha⁻¹ S and 30 t.ha⁻¹ manure.

According to Sá et al. 2015 and Stamford et al. 2015, the increase in Electrical

Conductivity with the doses of sulphur may have been due to the increase in the concentration of soluble ions such as Ca^{2+} , Mg^{2+} , $SO4^{-2}$ and H^+ generated by the oxidation of the S-element.

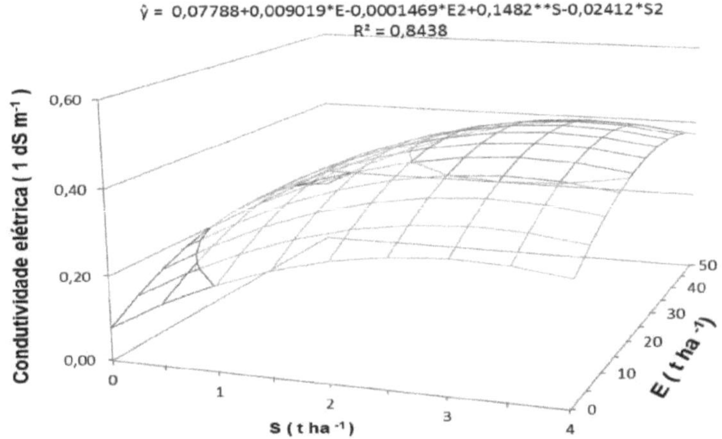

Figura 20 - **Response surface for Electrical Conductivity (EC) data at 30 days in response to S doses and bovine manure (IFNMG, Januária campus, 2016).**

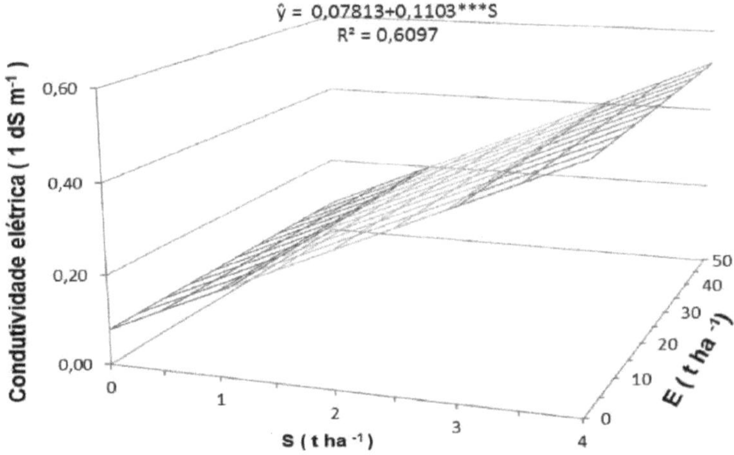

Figura 21 - **Response surface for Electrical Conductivity (EC) data at 60 days in response to S doses and bovine manure (IFNMG, Januária campus, 2016).**

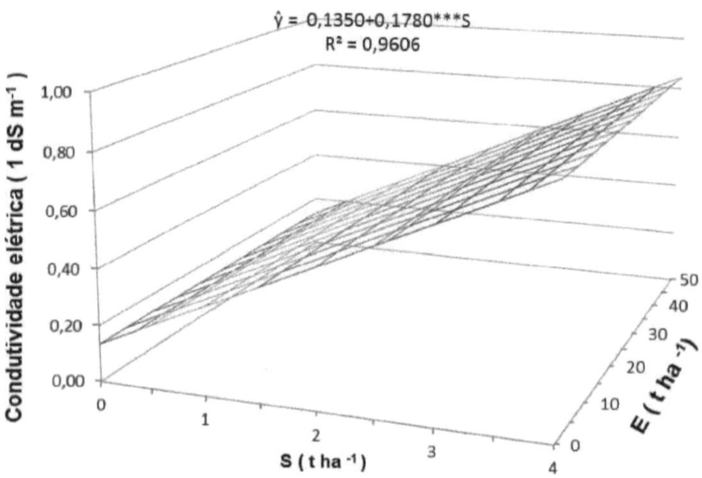

Figura 22   - Response surface for Electrical Conductivity (EC) data at 90 days in response to S doses and bovine manure (IFNMG, Januária campus, 2016).

When analysed at 184 days, the soil EC almost tripled, indicating an increase in the concentration of salts in the soil solution (Figure 23). The values went from 0.80 dSm- 1 in the absence of applications of S^0 and manure, to over 2.20 dSm-1 in the dose of 4 t ha-1 S^0 without the interference of manure. In the other analyses, at 214, 244 and 274 days, conductivity increased from 0.95, 0.76 and 0.73 dSm-1 to 1.80, 1.93 and 1.25 dSm-1 respectively (Figure 24, 25 and 26). The increase in electrical conductivity with the doses of sulphur may have been due to the increase in the concentration of soluble ions such as Ca^{2+} , Mg^{2+}, so_4^{-2} and H^+ itself produced by oxidising the S-element (SÁ et al. 2015, STAMFORD et al. 2015).

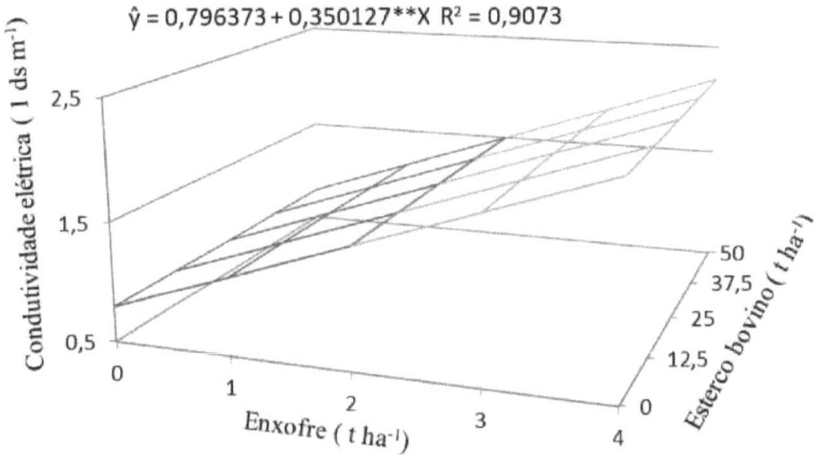

Figura 23   - **Response surface for Electrical Conductivity (EC) data at 184 days in response to S doses[0] and bovine manure (IFNMG, Januária campus, 2016).**

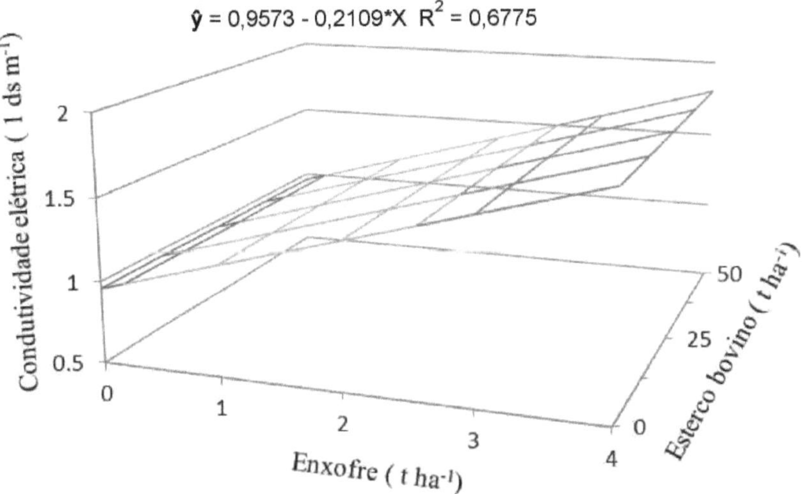

Figura 24   - **Response surface for Electrical Conductivity (EC) data at 214 days in response to S doses[0] and bovine manure (IFNMG, Januária campus, 2016).**

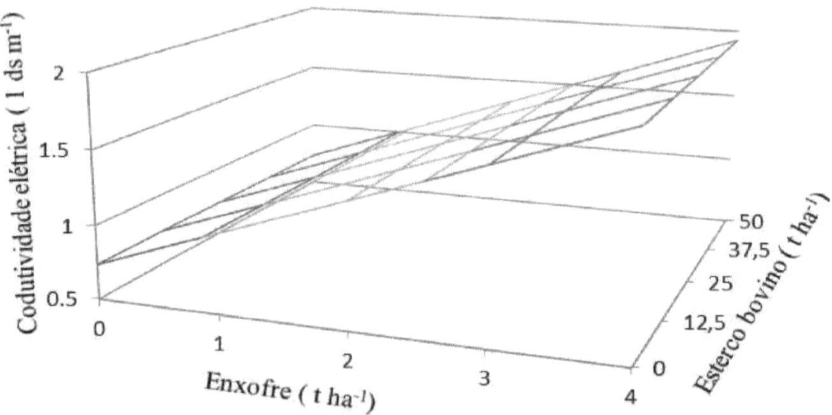

Figura 25   - Response surface for Electrical Conductivity (EC) data at 244 days in response to doses of S⁰ and bovine manure (IFNMG, Januária campus, 2016).

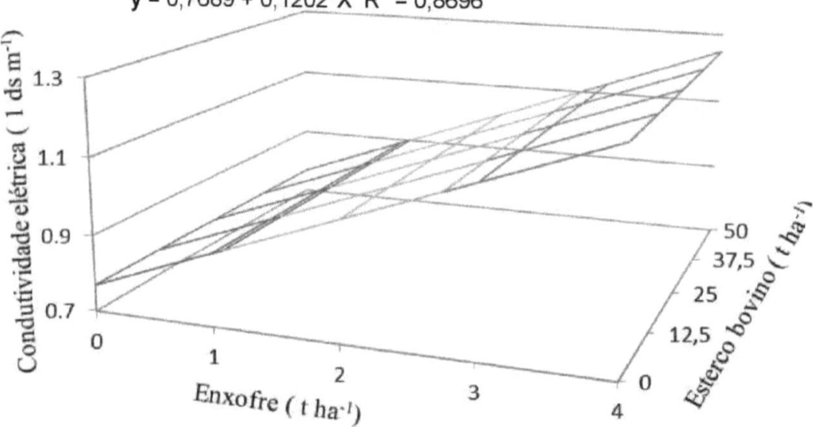

Figura 26   - Response surface for Electrical Conductivity (EC) data at 274 days in response to doses of S⁰ and bovine manure (IFNMG, Januária campus, 2016).

When this increase in soil EC is higher than that tolerated by the plant, its growth is directly jeopardised due to basically two processes: a reduction in water absorption resulting from the osmotic effect or water deficit; and a high concentration of ions in the transpiratory flow which causes damage to the leaves (MUNNS, 2005).

Calcium levels decreased in response to the doses of S^0 , with no interference from the doses of manure applied in the evaluation 184 days after the treatments were applied (Figure 27). The highest calcium value was 4.0 cmolc.dm^{-3} in the absence of S^0 and manure, and the lowest value was 2.02 cmol.dm^{-3} at the dose of 4 t ha^{-1} of S. Calcium concentrations at 214 days were not significantly influenced by the doses of S^0 and manure applied (Figure 28), and were estimated at an average of 4.13 cmol.dm^{-3} of Ca^{2+} regardless of the treatment.

$$\hat{y} = 3{,}9954 + 0{,}6570^{*}X - 0{,}1237^{*}X^2 \quad R^2 = 0{,}7336$$

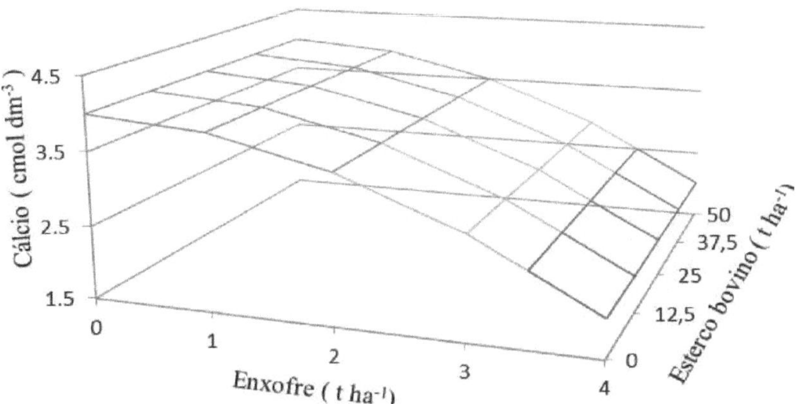

Figura 27 - **Response surface for calcium data at 184 days in response to doses of S^0 and cattle manure (IFNMG, Januária campus, 2016).**

$$\hat{y} = \bar{y} = 4{,}1279$$

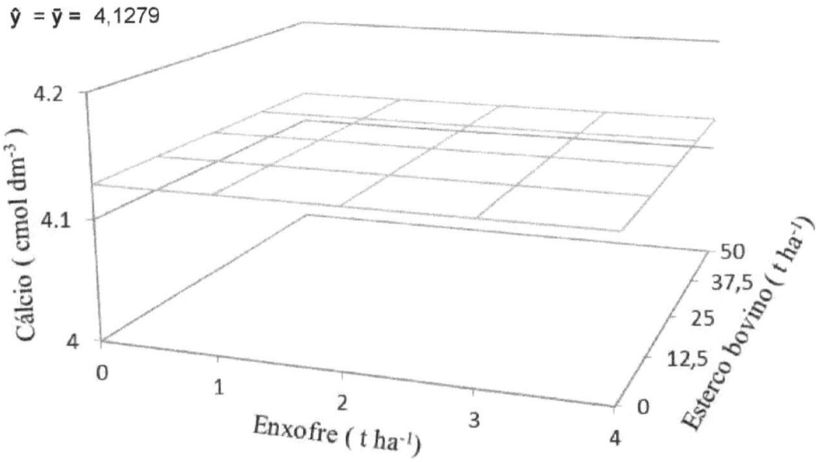

Figura 28   - Response surface for calcium data at 214 days in response to doses of S^0 and cattle manure (IFNMG, Januária campus, 2016).

At 244 days after the application of the treatments, the calcium values followed a quadratic response to the doses of S^0 and were not influenced by the application of manure (Figure 29). In the absence of S^0 and manure application, the value was 4.08 cmolc.dm^{-3} ; at the highest dose of 4.0 t ha^{-1} S^0 , the value was 4.22 cmolc.dm^{-3} of Ca^{2+} regardless of manure application. When assessed at 274 days after the treatments were applied, the calcium levels decreased in response to the doses of S^0 and were not affected by the application of manure (Figure 30). In the latter period, the highest value was 3.25 cmolc dm^{-3} of Ca^{2+} in the absence of S and manure, and the lowest estimate was 0.76 cmolc.dm^{-3} of Ca^{2+} at a dose of 4 t ha^{-1} S without manure. The monitoring of Ca^{2+} indicates that the period of 274 days after the application of the treatments was the one that allowed the greatest reduction in the concentration of calcium, indicating that the oxidation of S^0 , as well as reducing the pH, led to the formation of so_4^{2-} ionic pair responsible for the leaching of Ca^{2+} below the 20 cm depth of the soil.

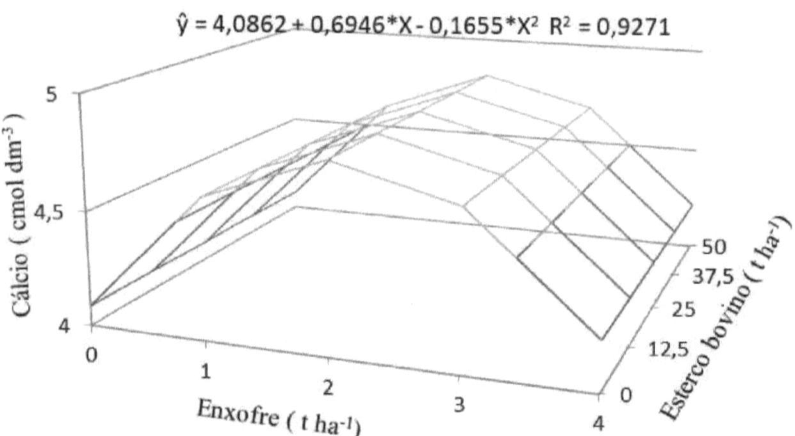

Figura 29   - Response surface for Calcium data at 244 days in response to S doses and cattle manure (IFNMG, Januária campus, 2016).

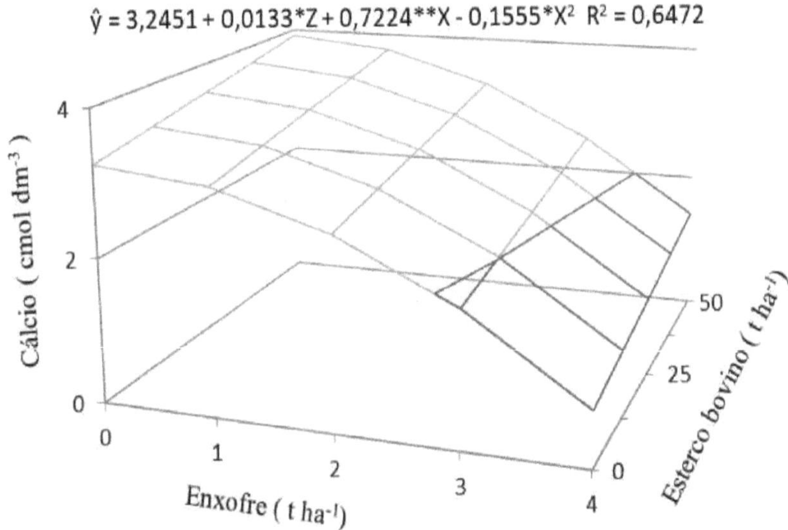

Figura 30 - **Response surface for calcium data at 274 days in response to doses of S⁰ and bovine manure (IFNMG, Januária campus, 2016).**

The magnesium values at 184, 214 and 244 days after the treatments were applied were not significantly affected by the doses of S^0 and manure (Figures 31, 32 and 33). At these times, all the values were below 1 cmolc.dm^{3-} of Mg^{2+}, indicating low availability of this nutrient in the soil. At 274 days, the concentrations of Mg^{2+} decreased in response to the doses of S^0 and manure (Figure 34), with the highest value being 0.28 cmolc.dm^{-3} in the absence of S^0 and manure, and the lowest 0.21 cmolc.dm^{-3} at the dose of 4 t ha^{-1} of S^0 regardless of the dose of manure.

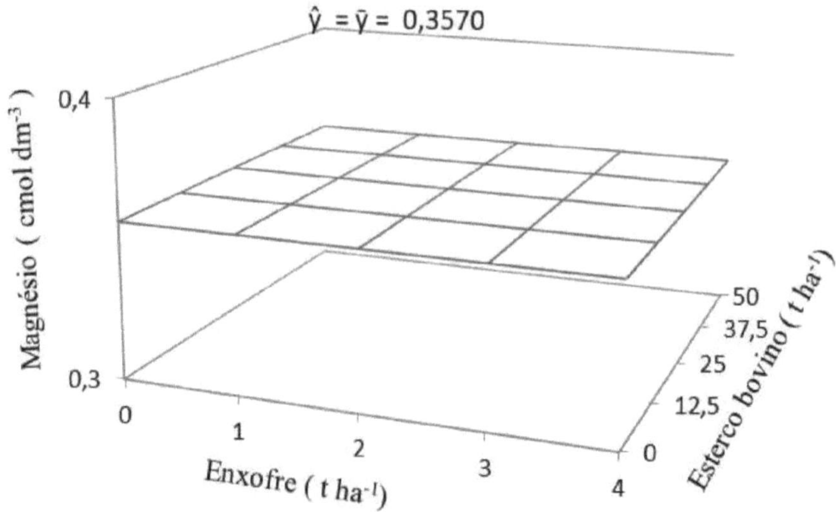

Figura 31 - **Response surface for magnesium data at 184 days in response to doses of S^0 and cattle manure (IFNMG, Januária campus, 2016).**

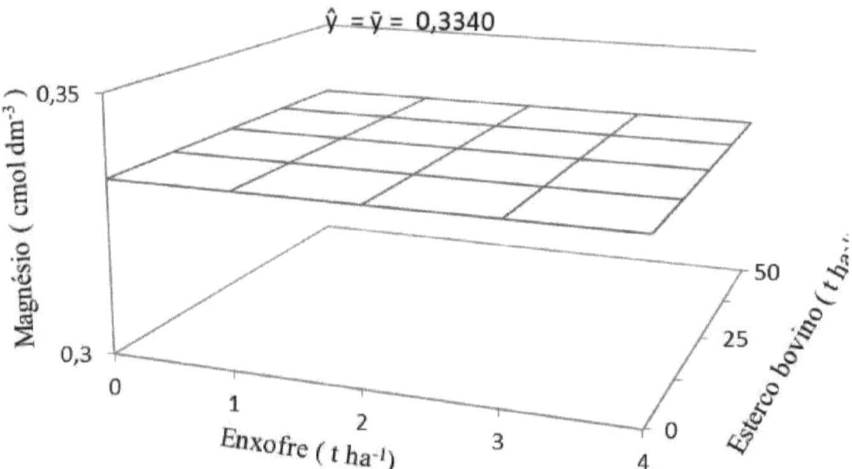

Figura 32 - **Response surface for magnesium data at 214 days in response to doses of S^0 and cattle manure (IFNMG, Januária campus, 2016).**

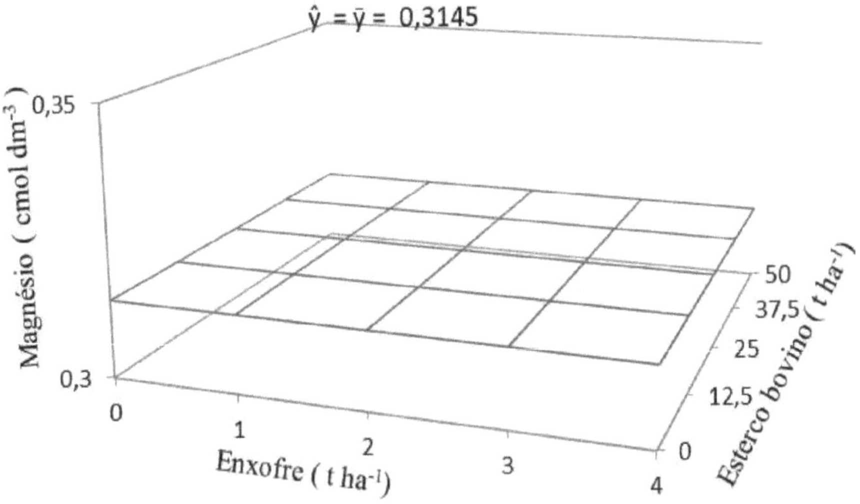

Figura 33 - **Response surface for magnesium data at 244 days in response to doses of S⁰ and cattle manure (IFNMG, Januária campus, 2016).**

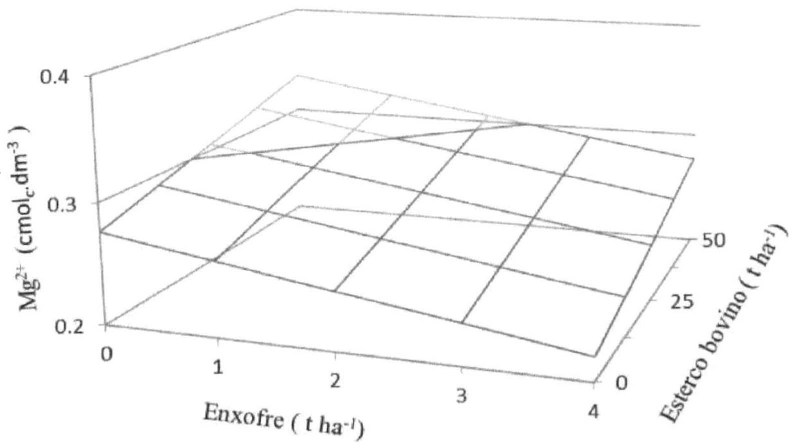

Figura 34 - **Response surface for magnesium data at 274 days in response to doses of S⁰ and cattle manure (IFNMG, Januária campus, 2016).**

Plants can be classified as halophytes, those that develop naturally in environments with high saline concentrations ranging from 50 to 500 mmol.dm-3 (ORCUTT & NILSEN, 2000), and glycophytes, those that show some kind of impairment of some phase of their cycle when the NaCl concentration exceeds 40 mmol.dm-3, equivalent to an electrical conductivity in the soil saturation extract of approximately 4.0 d.Sm-1 (RIBEIRO et al, 2007; MUNNS & TESTER, 2008). Thus, the use of S and manure in the soil, which allowed an increase of a maximum of 2.20 dsm-1, would not compromise the development of most crops.

CHAPTER 5

CONCLUSIONS

S^0 and cattle manure, especially the first source, raised the acidity of the soil, ensuring the correction of alkaline soils in the first 90 days of the experiment. However, only S0 increased acidity after the 90th day, guaranteeing the correction of alkaline soil during the 184th and 274th days of monitoring the experiment.

The best corrective effect on soil alkalinity occurred 30 days after the joint application of S^0 and cattle manure. Calcium values decreased, mainly in response to the reduction in pH promoted by the oxidation of $S .^0$

Suggesting the buffering effect of the soil, after 60 days of applying S^0 and cattle manure, the soil pH tended to rise again, without, however, reaching the initial alkaline value.

Without reaching critical limits to compromise plant growth, electrical conductivity increased in response to the application of S^0 and cattle manure, especially in response to the first source, attributing this effect to the increase in the concentrations of soluble ions Ca^{2+}, Mg^{2+}, $SO_4.^{-2}$

Electrical conductivity increased in response to the application of S^0, showing that its oxidation continued to promote an increase in the concentrations of soluble ions Ca^{2+}, Mg^{2+}, SO_4^{-2} throughout the 274-day period, but without reaching critical limits to compromise plant development.

CHAPTER 6

BIBLIOGRAPHICAL REFERENCES

AMARAL, F.C.S.; PEREIRA, N.R., CARVALHO JÚNIOR, W. Main limitations of Brazilian soils. Available at: Accessed on 18. June. 2015.

BARBER, S.A. Soil nutrient bioavailability: a mechanistic approach. 2ª ed. New York : J. Wiley, 1995. 414p. p.301-310. Chap.13: Sulfur.

BARROW, N. J. Slowly available sulphur fertilisers in south-western Australia. I. Elemental sulphur. Australian Journal of Experimental Agriculture and Animal Husbandry, Collingwood, v. 2, p. 211- 216, 1971.

BERNARDO, S.; SOARES, A. A.; MANTOVANI, E. C. Manual de irrigação. 8. ed. Viçosa: UFV, 2006. 625 p.

BISSANI, C.A.; TEDESCO, M.J. Sulphur in the soil. In: SYMPOSIUM ON SULPHUR AND MICRONUTRIENTS IN BRAZILIAN AGRICULTURE, 1988, Londrina, PR. Proceedings... Londrina: EMBRAPA. CNPSo :IAPAR : SBCS, 1988. p. 11- 29. Held during the XVII Brazilian Soil Fertility Meeting.

BRADY, N.C. Nature and properties of soils. 7. ed. Rio de Janeiro: Freitas Bastos, 1989. 898 p.

BURNS, G. R. Oxidation of sulphur in soils. Washington: The Sulphur Institute, 1967. 41 p. (Technical Bulletin, 13).

CHIEN, S. H.; FRIESEN, D. K.; HAMILTON, B. W. Effect of application method on availability of elemental sulfur in cropping sequences. Soil Science Society of American Journal, Madison, v. 52, p. 165-169, 1988.

CIFUENTES, F. R.; LINDEMANN, W. C. Organic matter stimulation of elemental sulfur oxidation in a calcareous soil. Soil Science Society of America Journal, Madison, v. 57, p. 727-731, 1993.

COELHO, F. S. Fertilidade do solo. 2. ed. Campinas: Instituto Campineiro de Ensino

Agrícola, 1973. 384 p.

COWELL, L. E.; SCHOENAU, J. J. Stimulation of elemental Sulphur oxidation by sewage sludge. Canadian Journal of Soil Science, Ottawa, v. 75, p. 247-249, 1995.

DENG, S.; DICK, R. P. Sulfur oxidation and rhodanese activity in soils. Soil Science, Baltimore, v. 150, n. 2, p. 552-560, 1990.

DUKE, S. H.; REISENAUER, H. M. Roles and requirements of sulfur in plant nutrition. In: SULFUR in agriculture. Madison: ASA/ CSSA/SSSA, 1986. p. 123-168. (Agronomy monography, 27).

DUKE, S.H.; REISENAUER, H.M.. Roles and requirements of sulfur in plant nutrition. In: SULFUR in agriculture. Madison : ASA : CSSA : SSSA, 1986. p. 123-168. (Agronomy monography, 27).

EMBRAPA. National Soil Survey and Conservation Service (Rio de Janeiro, RJ). Manual of soil analysis methods. 2.ed. Rio de Janeiro, 1997. 212p.

FOX, R. L.; ATESALP, H. M.; KAMPBELL, D. H.; RHOADES, H. F. Factors influencing the availability of sulfur fertilizers to alfalfa and corn. Soil Science Society Proceedings, Madison, p. 406-408, 1964.

FRENEY, J.R. Forms and reactions of organic sulfur compounds in soils. In: SULFUR in agriculture. Madison : ASA : CSSA : SSSA, 1986. p. 207-232. (Agronomy monography, 27).

GERMIDA, J.J.; JANZEN, H.H. Factors affecting the oxidation of elemental sulfur in soils. Fertiliser Research, Wageningen, Netherlands, v.35, p.101-114, 1993.

GRAYSTON, S.J.; GERMIDA, J.J. Influence of crop rhizospheres on populations and activity of heterotrophic sulfur-oxidising microorganisms. Soil Biology & Biochemistry, Oxford, UK, v.22, n.4, p.457-463, 1990.

HASHEMIMAJD, K.; FARANI, T. M.; SOMARIN, S. J. Effect of elemental sulphur and compost on pH, electrical conductivity and phosphorus availability of one clay soil. African Journal of Biotechnology, v. 11, n.6, p. 1425-1432, 2012.

HEYDARNEZHAD, F. et al. Influence of elemental sulfur and sulfur oxidizing bacteria on some nutrient deficiency in calcareous soils. International Journal of Agriculture

and Crop Sciences, v.4, n.12, p.735-739, 2012. Available at: <http://ijagcs.com/wp-content/uploads/2012/09/735-739.pdf>. Accessed on: 15 June 2015.

HOROWITZ, N. Oxidation and agronomic efficiency of elemental sulphur in Brazilian soils. 2003. Thesis (PhD) - Postgraduate Programme in Soil Science, Faculty of Agronomy, Federal University of Rio Grande do Sul, Porto Alegre, 2003.

JANZEN, H. H.; BETTANY, J. R. Oxidation of elemental sulfur under field conditions in Central Saskatchewan. Canadian Journal of Soil Science, Ottawa, v. 67, p. 609618, 1987b.

JANZEN, H. H.; BETTANY, J. R. The effect of temperature and water potential on sulfur oxidation in soils. Soil Science, Baltimore, v. 144, n. 2, p. 81-89, 1987c.

JANZEN, H.H. Elemental sulfur oxidation as influenced by plant growth and. degree of dispersion within soil. Canadian Journal of Soil Science, Ottawa, v.70, p.499-502, 1990.

JANZEN, H.H.; BETTANY, J.R. Measurement of sulfur oxidation in soils. Soil Science, Baltimore, v. 143, n.6, p.444-452, 1987a.

LAWRENCE, J.R.; GERMIDA, J.J. Relationship between microbial biomass and elemental sulfur oxidation in agricultural soils. Soil Science Society of American Journal, Madison, v.52, p.672-677, 1988.

LEE, A.; WATKINSON, J. H.; ORBELL, G.; BAGYARAJ, J.; LAUREN, D. R. Factors influencing dissolution of phosphate rock and oxidation of elemental sulphur in some New Zealand soils. New Zealand Journal of Agricultural Research, Wellington, v. 30, n. 3, p. 373-385, 1987.

International Handbook of Soil Fertility / translated and adapted by Alfredo Scheid. -

2ed., revised and expanded. - Piracicaba: POTAFOS, 1998 ,177 p. Il.

MAGALHÃES, R. et al. Use of sulphur to correct the pH of organic compost. In: SALÃO DE INICIAÇÃO CIENTÍFICA UFRGS, 13, 2005, Porto Alegre, RS. **Proceedings...** Porto Alegre: ROPESQ, 2005. CD-Resumes.

McCASKILL, M. R.; BLAIR, G. J. Particle size and soil texture effects on elemental sulfur oxidation. Agronomy Journal, Madison, v. 79, p. 1079-1083, 1987.

MOHAMED, A. I.; ALI, O. M.; MATLOUB, M. A. Effect of soil amendments on some physical and chemical properties of some soils of Egypt under saline irrigation water. African Crop Science Conference Proceedings, El-Minia, v. 8, n.1 , p. 1571-1578, 2007.

MUNNS, R. 2005. Genes and salt tolerance: bringing them together. New Phytologist, v.167, n.3, p.645-663.

MUNNS, R.; TESTER, M. 2008. Mechanisms of salinity tolerance. Annu. Rev. Plant Biol. v.59, p.651-681.

NEPTUNE, A.M.L.; TABATABAI, M.A.; HANWAY, J.J. Sulfur fractions and carbon- nitrogen-phosphorus-sulfur relationships in some Brazilian and Iowa soils. Soil Science Society of America Proceedings, Madison, v.39, p. 51-55, 1975.

NOR, Y.M.; TABATABAI, M.A. Oxidation of elemental sulfur in soils. Soil Science Society of America Journal, Madison, v.41, p. 736-741, 1977.

ORMAN,S.;KAPLAN,M.Effects of elemental sulfur and farmyard manure on pH and salinity of calcareous sandy loam soil and some nutrient elements in tomato plant. Journal of Agricultural Science and Technology,v.5,p.20-26,2011

PEREIRA, E. B. et al. Performance of correctives in the recovery of a saline-sodic soil in the semi-arid region of Paraiba. In: REUNIÃO BRASILEIRA DE FERTILIDADE DO SOLO E NUTRIÇÃO DE PLANTAS, 29, Guarapari, 2010. Proceedings... Guarapari: Brazilian Society of Soil Science, 2010. 1 CD-ROM.

REHM, G. W.; CALDWELL, A. C. Sulfur supplying capacity of soils and the relationship to soil type. Soil Science, Baltimore, v. 105, n. 5, p. 355-361, 1968.

RIBEIRO, A.C.; GUIMARÃES, P.T.G. & ALVAREZ V., V.H. (ed.). Recommendation for the use of correctives and fertilisers in Minas Gerais: 5ª approximation. Viçosa: CFSEMG, 1999. 359p.

RIBEIRO, J. S.; LIMA, A. B.; CUNHA, P. C.; WILLADINO, L.; CÂMARA, T. R. O. 2007. Abiotic Stress in Semi-Arid Regions: Metabolic Responses of Plants. In: MOURA, A. N.; ARAUJO, E. L.; ALBUQUERQUE, U. P. (eds.) Biodiversity, economic potential and ecophysiological processes in northeastern ecosystems, Recife: Comunigraf. 361p.

RIBEIRO, M. S.; LIMA, L. A.; FARIA F. H. S.; SANTOS, S. R.; KOBAYASHI, M. K. Classification of water from tube wells in the north of the state of minas gerais for irrigation. Engenharia na Agricultura, v.18, n.3, p.17, 2010.

SÁ, F. V. S. et al. Initial growth of craibeira in salinised soil corrected with elemental sulphur. Irriga, Botucatu, v. 18, n. 4, p. 647-660, 2013.

SÁ, F. V. S. et al. Influence of gypsum and biofertiliser on the chemical attributes of a saline-sodic soil and on the initial growth of sunflower. Irriga, Botucatu, v. 20, n. 1, p. 46-59, 2015.

SAEG. SAEG: system for statistical analyses, version 9.1. Viçosa: UFV, 2007.

SAIK, R.D. The evolution of a sulphur bentonite fertiliser: one company's perspective. Sulphur in Agriculture, Washington, v.19, p.74-77, 1995.

SIERRA, C.B. et al. Azufre elemental como corrector del pH y la fertilidad de algunos suelos de la III y IV región de chile. Agricultura Técnica, v.67,n.2, p.173-181, 2007. Available at: <http://dx.doi.org/10.4067/S0365- 28072007000200007>. Accessed on: 02 June. 2015. doi: 10.4067/ S0365-28072007000200007

SKIBA, U.; WAINWRIGHT, M. Oxidation of elemental-S in coastal-dune sands and soils. Plant and Soil, Dordrecht, Netherlands, v.77, p.87-95, 1984.

SOUSA, F. Q. et al. Growth and physiological responses of tree species in salinised soil treated with correctives. Revista Brasileira de Engenharia Agrícola e Ambiental, Campina Grande, v. 16, n. 2, p. 173-181, 2012.

STAMFORD, N. P. et al. Effect of sulphur inoculated with Thiobacillus on saline soils amendment and growth of cowpea and yam bean legumes. Journal of Agricultural Science, Toronto, v. 139, n. 3, p. 275 - 281, 2002.

STAMFORD, N. P.; RIBEIRO, M. R.; CUNHA, K. P. V. Effectiveness of sulfur with Acidithiobacillus and gypsum in chemical attributes of a Brazilian sodic soil. World Journal Microbiology Biotechnology, Netherlands, v. 23, n. 10, p. 1433-1439, 2007.

STAMFORD, N. P. et al. Effect of gypsum and sulfur with Acidithiobacillus on soil salinity alleviation and on cowpea biomass and nutrient status as affected by PK rock biofertilizer. Scientia Horticulturae, Amsterdam, v. 192, n. 1, p. 287-292, 2015.

TEDESCO, M.J.; GIANELLO, C.;BISSANI, C.A.; BOHNEN, H.; VOLKWEISS, S.J.; Analyses of soil, plants and other materials. 2 ed. rev. and ampl. Porto Alegre: UFRGS Soil Department, 1995. 174p. (Technical Bulletin of Soils).

TISDALE, S.L.; NELSON, W.L.; BEATON, J.D.; HAVLIN, J.L. Soil fertility and fertilisers. 5. ed. New York : MacMillan, 1993. 634p. p.266-303. Chap.8: Soil and fertiliser sulphur, calcium and magnesium.

WAINWRIGHT, M. Sulfur oxidation in soils. Advances in Agronomy, San Diego, CA, v.37, p.349-396, 1984.

WAINWRIGHT, M.; KILLHAM, K. Microbial transformations of some particulate pollution deposits in soils-A source of plant-available nitrogen and sulphur. Plant and Soil, Dordrecht, Netherlands, v.65, p.297-301, 1982.

WAINWRIGHT, M.; KILLHAM, K. Sulphur oxidation by fusarium solani. Soil Biology & Biochemistry, Oxford, UK, v.12, p.555-558, 1980.

WAINWRIGHT, M.; NEVELL, W.; GRAYSTON, S. J. Effects of organic matter on sulphur oxidation in soil and influence of Sulphur oxidation on soil nitrification. Plant and Soil, Dordrecht, v. 96, p. 369-376, 1986.

WATKINSON, J.H. Measurement of the oxidation rate of elemental sulfur in soil. Australian Journal of Soil Research, Collingwood, v.27, p. 365-375, 1989.

MCCASKILL, M.R.; BLAIR, G.J. Particle size and soil texture effects on elemental sulfur oxidation. Agronomy Journal, Madison, v.79, p.1079- 1083, 1987.

WATKINSON, J.H.; BLAIR, G.J. Modelling the oxidation of elemental sulphur in soils. Fertiliser Research, Netherlands, v.35, p.115-126, 1993.

More
Books!

info@omniscriptum.com
www.omniscriptum.com
OMNIScriptum

Printed by Books on Demand GmbH, Norderstedt / Germany